MONITORING THE IRAQ/KUWAIT CONFLICT

2nd Edition, November, 1990

1st Reprint

Published by

Interbooks, 8 Abbot Street, Perth, PH2 0EB, Scotland

CONTENTS

INTRODUCTION

This is now the second edition of this publication, which has been prompted by a wealth of information that came to light after publication of the first edition, and by the great success in sales which saw the first edition being sold out in Europe only weeks after its printing.

This edition has been complied over the first three months of the conflict, and in that time has seen desert stations grow from hastily constructed stations to impressive multi-station networks.

Each chapter has in turn been re-researched, with all relevant information being updated. Frequency lists have been increased, and new sections on channel designators and callsigns, as well as the US Army have been included. Additional UHF satellite information has also been added for those listeners interested in the UHF spectrum.

As in the first edition, all topics are covered: Fighter pilots talking to ground stations, Air Force One talking to the State Department, the US Navy in the Gulf, cargo vessels' ship to shore radio and broadcasting stations. The most active frequencies remain underlined.

In order to keep entries as accurate and as up-to-date as possible, readers are invited to send their frequencies and comments to Interbooks, 8 Abbot Street, Perth, PH2 0EB, Scotland, where they will be passed on to the author and will be included in any future edition.

I would also like to take this opportunity to thank all those readers who have taken the time to write letters of encouragement and have supplied me with priceless information. Your letters are always welcome.

Langley Pierce,
November, 1990

AIR FORCE

US AIR FORCE

Mode:USB

The logistics of Operation Desert Shield are being carried out almost entirely by the US Air Force Military Airlift Command (MAC). Flights of C-141 Starlifters and giant C-5 Galaxy aircraft are flying into the Gulf round the clock, arriving at Saudi bases every 10 minutes! Gulf bound MAC flights can be distinguished from other traffic as they all have three digit call-signs suffixed by either SK or SM. More recently, Gulf bound aircraft are using four digit mission numbers prefixed by a single letter.

Once airborne from their stateside bases, aircraft will first come into military HF range over mid Atlantic when they begin calling European ground stations, such as Lajes in the Azores, or Croughton in the United Kingdom. Once into the Mediterranean region, they switch to the station at Incirlik in Turkey. And on leaving the Mediterranean on their final leg, aircraft can be heard calling "Dhahran MAC Ops" on 11176 kHz.

Traffic to these stations consists of weather requests and ETA's to refuelling stops at Zaragoza in Spain, or Rhein Main in West Germany, and ETA's to their destinations in Saudi or the United Arab Emirates. Air bases are usually identified by their ICAO designator rather than by name:

EDAF - Rhein Main	OEKM - Khamir Mushait
LEZA - Zaragoza	OERY - Riyadh
OEDR - Dhahran	

Global Control & Communications Stations (GCCS):

RAF Croughton, United Kingdom — Mode: USB

3067 kHz	9011	15036
5703	11176	
6750	13214	

Croughton is also using 11180.0 kHz as a MAC discrete frequency.

Lajes Air Base, Azores Mode: USB

3081 kHz	8967	13244
4746	11271	
6750	13214	

Incirlik Air Base, Turkey Mode: USB

6738 kHz	15015
11176	15036
13214	23227

Dhahran MAC Ops, Saudi Arabia Mode: USB

11176 kHz

In the search for a diplomatic solution to the crisis, or to simply boost moral, various US Government officials, such as Secretary of State James Baker, and National Security Adviser Dick Chaney have made trips to Europe and the Gulf. These VIP Special Air Mission flights, characterised by their call-sign SAM, can often be heard on GCCS channels, but are more usually heard on discrete Mystic Star (Foxtrot Listings) channels working Andrews AFB in Maryland.
Mystic Star channels active during the crisis so far are given below:

Frequency	Designator	Flight
6756 kHz	F151	SAM 971 (James Baker)
11058	F214	SAM 127
11118	F455	SAM 127
11226	F535	Air Force One and SAM 974
13205	F474	Air Force One
13217	F537	Air Force One
13823	F377	Air Force One and SAM 974
14829	?	SAM 973 (Dick Chaney)
18320	F68	Air Force One and SAM 27000
23385	F479	SAM 972 (Dick Chaney)

All these flights used upper side band, though the Mystic Star Network can also operate in lower side band. Other frequencies in the Mystic Star Channel Plan include:

5710 kHz	6790	7858	9120	11052	11466	17972
5760	6812	7997	9270	11055	11484	18175
5800	6817	8040	9320	11056	11488	18532
5820	6830	8162	9414	11058	11498	23035
6683	6918	9017	10112	11118	13823	
6716	6927	9023	10427	11156	13878	
6730	6993	9043	10583	11220	15015	
6756	7813	9057	11035	11407	15680	

During SAM flights, some of these frequencies carry encrypted and plain text teletype to and from the aircraft. The older aircraft use 75 baud standard RTTY, while the newer aircraft use 300 baud ASCII.

Secret Service callsigns that are often heard on these frequencies include:

Cactus – Camp David
Carpet – White House Garage
Crown – White House Communications Centre
Royal Crown – High Priority White House Comms
Snowbank – Barbara Bush
Timberwolf – President Bush

SAM flights also use AFSATCOM (Air Force Satellite Communications) which is the primary communications relay method for the USAF. The satellite based system is shared with the US Navy and provides UHF communications links anywhere in the world, though USAF Gulf activity has yet to be heard.

Confirmed voice frequencies include: Mode: FM

261.500 MHz	261.950	262.300
261.600	262.050	262.475
261.650	262.100	263.675
261.750	262.225	

Another satellite link that is often referred to is the Orderwire or Combat Cinders. This system allows for secure voice, teletype and fax transmissions in wide FM, LSB, USB or narrow AM modes. It is used extensively by both SAM and Air Force One/Air Force Two flights. Traffic is most likely to be heard not on the designated frequencies but rather 50kHz either side of the central carrier, so it may be worth tuning around to see what is there.

Channel	Ground	Air	
RF1	326.000 MHz	382.350 MHz	
RF2	246.950	305.550	
RF3	345.500	336.800	
RF4	322.750	366.000	
RF5	390.000	397.050	
RF6	382.350	326.000	RF1 Reverse
RF7	305.550	246.950	RF2 Reverse
RF8	336.800	345.500	RF3 Reverse
RF9	366.000	322.750	RF4 Reverse
RF10	397.050	390.000	RF5 Reverse

Communications channels of the Strategic Air Command (SAC), the USAF's fighting wing, will almost certainly come alive if the conflict escalates. SAC maintains an HF communications network called Giant Talk, which is active every day of the year with routine tactical traffic. But this traffic would most certainly be joined by air-to-ground communications from F-111 bombers from Incirlik or B-52 heavy bombers from Diego Garcia on their way to targets in Iraq.

SAC Channels: Mode: USB

Frequency	Designator	
4725 kHz	S-390	
6761	S-391	Secondary A-A refuelling.
9027	S-392	
11243	S-393	Primary Day Channel.
13241	S-394	
17975	S-395	

SAC also maintained several tactical nets used by Airborne Command Post aircraft. These frequencies include:

PAPA 38 Series: Mode: USB

P-383 15044 kHz

SEIRRA 30 Series: Mode: USB

S-302	3113 kHz
S-303	3295
S-304	4495
S-309	9057
S-310	11220
S-311	11494
S-312	13211

WHISKEY 10 Series: Mode: USB

W-101	5800 kHz
W-103	6757
W-104	7475
W-105	7831
W-108	12070
W-109	13247
W-111	17972
W-116	20167

SAC also used to maintain other frequencies each designated by a one or two letter group, so its worth tuning around these frequencies for activity. The most prominent discrete frequencies today are all taken from this out-dated bandplan of the early 80's. Nonetheless, these designators may come in handy, as has already been demonstrated by some die hard pilots who prefer to use the old designator system!

Mode: USB

Designator	Frequency	Designator	Frequency
A	11243 kHz	FS	21815 kHz
AC	13907	FX	6680
AM	3295	G	6826
AP	8101	I	15962
AS	3369	J	18046
AT	14744	K	6870
AZ	11607	L	11494
B	11220	M	15041
BH	17617	O	6863
BQ	5700	P	9057
BU	5826	Q	6761
BW	13211	R	9027

BX	15091	S	13241
C	14955	SE	14716
CA	20846	T	17975
CQ	15035	U	23337
D	20890	V	4725
E	4495	W	20631
F	5020	X	7330
FC	7983	Y	7330
FQ	5684	ZQ	11408

For those readers who are UHF satellite capable, the worldwide SAC alert frequency is channel 9 on the AFSATCOM, tentatively identified as 244.090MHz, with the SAC Missile Command channel on 396.250MHz

The AFSATCOM bandplans consists of:

Mode: FM

Chan No.	Pacific Ocean Relay -Alpha-	Atlantic Ocean Relay -Bravo-	Indian Ocean Relay -Charlie-
12	243.945 MHz	244.045MHz	244.145MHz
13	243.955	244.055	244.155
14	243.960	244.060	244.160
15	243.965	244.065	244.165
16	243.970	244.070	244.170
17	243.975	244.075	244.175
18	243.980	244.080	244.180
19	243.985	244.085	244.185
20	243.990	244.090	244.190
21	243.995	244.095	244.195
22	244.000	244.100	244.200
23	244.010	244.110	244.210

Many of these channels carry only encrypted RTTY and other digital data modes, with very few being used for voice.

Air Force MARS activity in the Gulf is now being coordinated by two USAF wings, the Tactical Air Command and the Military Airlift Command. Two stateside TAC stations were initially operating two stations in Saudi, one possibly in Dhahran. Stations in the net were:

Fireside 01	Langley AFB, Virginia
Fireside 03	Shaw AFB, S. Carolina
Fireside 11	Saudi Arabia
Fireside 12	Saudi Arabia

Around the middle of September, 1990, the net dramatically altered its appearance. The previously used callsigns reverted back to the widely accepted USAF MARS format.

Present	Previously
AGA2LA	Fireside 01
AGA2SH	Fireside 03
AFQ371	Fireside 11

MAC operates the second net, linking Gulf stations, European stations, and quite often aircraft in flight. Both nets can be found in USB using two separate channel plans:

Alpha to Papa Channel Designators:

Frequency	Designator
11121.0	A
11407.0	B
13927.0	D
14392.0	E
14405.0	F
14607.5	G
17487.0	J
20188.5	M
20737.0	O
20992.5	P

Alpha Charlie Channel Designators:

Frequency	Designator
11407.0 kHz	ACA
13927.0	ACB
13977.0	ACC
14390.5	ACE
14606.0	ACF
20188.5	ACG

Note that many of the frequencies are duplicated in either listing. ACF, 14606 kHz, is the calling channel, where most of the activity can be heard. Net security is high, with channels being referred to only by designator, and never by frequency. Additionally, the unusual frequency of 6996 kHz, USB, is often active in the early hours of the morning discussing propagation conditions into the Gulf area for the following day.

Callsigns in the net include:

AFA1MH	Stateside
AFA1YV	Stateside
AFA2DE	Stateside
AFA2KD	Stateside
AFA2SP	Stateside
AFA2XD	Stateside
AFA3BZ	Stateside
AFA4KE	Kelly AFB, Texas
AFA7BQ	Europe
AFA7DW	Europe
AFA7MJ	Ramstein AB, Germany
AFA7NL	Europe
AFA7TJ	Europe
AFB2DE	Ft Lauderdale, Florida
AFB2RF	Stateside
AFF2SP	Miami, Florida
AFQ375	Saudi Arabia
AFQ376	Saudi Arabia
AFQ379	Saudi Arabia
AFQ472	Saudi Arabia
AFT7MP	Europe
AGA1MC	MacGuire AFB, New Jersey
AGA2MB	Myrtle Beach AFB, S. Carolina
AGA2PA	Patrick AFB, Florida
AGA2PO	Pope AFB, North Carolina
AGA2PX	Deployed Unit in Saudi Arabia
AGA3HQ	Scott AFB, Illinois
AGA7BI	Bitburg AB, Germany
AGA7DW	Europe
AGA7IR	Iraklion AB, Greece
AGA7LI	Lindsey AB, Germany
AGA7MJ	Ramstein, Germany
AGA7ML	Europe

AGA7RM	Rhein Main AB, Germany
AIR	USAF MARS HQ, Andrews AFB
CUW20	Net Coordinator, Lajes AB, Azores
ENN1183	Saudi Arabia

It is quite likely that US Army stations will be heard on this net as there is some confusion among units as to which net that they are allowed to operate on.

MARSgrams can be heard on many of these frequencies in voice or RTTY. The RTTY data carrier is 300Hz above the USB voice carrier and is sent at 75 baud with a 170Hz shift in reverse sense.

The typical MARSgram format consists of:

```
DE AGA7RM MSG NR 2004
R 270916Z SEP 90
FM     (Name and address of sender)
TO     (Name, address and telephone number of
        recipient)
GR 25 (Word count in message)
BT
       (Message)
BT
CFN:   (Name, address and telephone number of
        recipient)
NNNN
```

ROYAL AIR FORCE Mode: USB

The RAF communications effort in the Gulf is very limited indeed, with their ground attack aircraft relying almost entirely on UHF frequencies, which is out of the range of all but those who are in the area. Alternatively, the RAF's Support Command can be heard working two of the RAF's ground control stations, Architect, at RAF Upavon, and Cyprus Flight Watch, at RAF Akrotiri in Cyprus.

Support Command flights are characterised by their ASCOT call-signs, and can be heard working Architect and Cyprus Flight Watch with ETA's, SELCAL checks and routing information. Flights depart from RAF Brize Norton, RAF Lyneham and RAF Wittering, and report at 15° East before continuing the final leg of their journey into Akrotiri. Several flights continue on after Akrotiri to Saudi bases with supplies for RAF aircraft stationed in the Gulf.

The RAF current maintains three bases in the Gulf. The first at Dhahran operates Tornado GF1 Fighter aircraft. The second is in Bahrain and operates a squadron of Jaguar ground attack aircraft, while the third at an undisclosed location in Saudi operates the ground attack version of the Tornado.

Architect, RAF Upavon, United Kingdom Mode: USB

4742 kHz	9032	18018
5713	11204	23220
6738	11234	

Cyprus Flight Watch, Akrotiri, Cyprus. Mode: USB

4730 kHz	11234
9032	18018
11204	

ROYAL SAUDI AIR FORCE Mode: USB

The Royal Saudi Air Force maintains several air bases within the country. Principal sites are located at:

Dhahran	King Khalid City
Khamir Mushait	Nejran

These bases are presently being shared with RAF and USAF aircraft, with Saudi Tornados flying sorties alongside them. Communications are believed to be carried out on UHF with coordination AWACS aircraft, but some HF activity is expected.

3095 kHz
8967
11243

NAVY

US NAVY

Mode: USB

The US Navy, like the USAF, utilizes both HF and UHF channels in its worldwide communications. The secondary method, and most accessible to listeners are the HF bands, the primary methods being UHF satellite communications.

US Navy HICOM (High Command) Channels:

	Atlantic Ocean	Indian Ocean
Primary	6720 kHz	23315 kHz
Secondary	11255 kHz	12215 kHz
Tertiary	23287 kHz	

These channels are used primarily for passing tactical traffic, and are arranged in a simulcast structure, so that the listener can choose from the two or three channels for the best reception.

Stations operating in these net adopt tactical alphanumeric call-signs such as 'One Kilo Foxtrot' or 'Alpha Nine India.' Traffic is encrypted, though not voice scrambled, and the casual listener will be able to hear stations setting up and coordinating ship-to-ship and ship-to-shore communications data links.

During major emergencies, such as Hurricane Hugo which hit the Caribbean in 1989, evacuation and rescue communications were often heard on HICOM channels in plain language. In a similar incident, the fire on board the US Navy battleship Iowa in that same year, it was the HICOM channels which first broke the news!

Playing just as an important role are the often unlisted Navy tactical channels. These channels are used in tactical exercises and operations to coordinate activities in operational locations. Although not as active as the HICOM channels, and prone to sporadic use, a wealth of interesting and informative communications can be heard in USB on these channels:

4702 kHz	6804
4711	11191
6770	11205

Moving on to the fleets primary mode of communication, the FLTSATCOM (Fleet Satellite Communications) consists of five or so satellites in geosynchronous orbit stationed over the Atlantic, Pacific and Indian Oceans, and are able to relay traffic sent from shore transmit stations direct to Naval vessels anywhere in the world. Operating in UHF bands, communications are generally in the form of encrypted data, although several voice channels are known.

Mode: FM

Chan No.	Pacific Ocean Relay -Alpha-	Atlantic Ocean Relay -Bravo-	Indian Ocean Relay -Charlie-
1	250.250MHz	250.550MHz	250.650MHz
2	260.600	261.700	262.300
3	251.950	252.050	252.150
4	253.650	253.750	253.850
5	255.350	255.450	255.550
6	256.950	257.050	257.150
7	258.450	258.550	258.650
8	265.350	265.450	265.550
9	266.850	266.950	267.050
10	268.250	268.350	268.450
11	269.750	269.850	269.950

Note that channel 2 is a 500kHz wideband channel. The frequency quoted is the central frequency. Other frequencies include:

Mode: FM

MHz	Chan No.	Bandplan	Usage
261.475	2	Y	Atlantic Relay
261.575	6	Y	Fleet Relay
261.800	15	Y	Navy Call Channel
262.150	5	Z	Pacific Relay
262.500	19	Z	Fleet Relay
262.550	-	-	Pacific Relay

One of two unique moral boosting telephone services (the other being that of the Canadian Forces) operated by the US Military is the MARS (Military Affiliate Radio System) Network. The Navy MARS operations are carried out in USB on a variety of frequencies in the HF bands. The Calling Channel, where all ships call first for a shore station is 14441.5 kHz.

13643	14383.5	14470	14818.5	20936
13826	14463.5	14477	16387.5	
13974	14467	14483.5	20678.5	

The MARS network operates a static call-sign system whereby the listener can identify the ship that he or she is listening to, although ships are forbidden by COMSEC regulations to identify themselves by name. Below is a list of US Navy ships along with their MARS callsigns which have been heard, or are known to be in the Gulf Area. I have included hull numbers so that ships may also be identified on news reports.

Ships:

Call	Ships Name	Type	Hull No.
NNNOCAI	USS Antietam	CG	54
NNNOCBD	USS Thomas S. Gates	CG	51
NNNOCJN	USS Reid	FFG	30
NNNOCMU	USS Seattle	AOE	3
NNNOCNE	USS England	CG	22
NNNOCOA	USS Raleigh	LPD	1
NNNOCOE	USS Independence	CV	62
NNNOCOO	USS Tattnal	DDG	19
NNNOCOQ	USS Pensacola	LSD	38
NNNOCOU	USS Saratoga	CV	60
NNNOCSB	USS Spartanburg County	LST	1192
NNNOCTQ	USS O'Brien	DD	975
NNNOCTN	USS Barbey	FF	1088
NNNOCUX	USS Nassau	LAH	4
NNNOCVA	USS Belknap	CG	26
NNNOCVG	USS Dwight D. Eisenhower	CVN	69
NNNOCVU	USS Saginaw	LST	1186
NNNOCWI	USS Wisconsin	BB	64
NNNOCWO	USS Flint	AE	32
NNNOCXB	USS John F. Kennedy	CV	67

NNN0CXN	USS Portland	LSD	37
NNN0NXK	USS Iwo Jima	LPH	2
NNN0NZR	USS La Salle	AGF	3

AE	Ammunition Ship
AGF	Command Flag Ship
AOE	Combat Support Ship
BB	Battleship
CG	Guided Missile Cruiser
CV	Aircraft Carrier
CVN	Nuclear Aircraft Carrier
FF	Frigate
FFG	Guided Missile Frigate
LAH	Landing Craft
LPD	Amphibious Transport Dock
LPH	Amphibious Craft
LSD	Landing Ship, Dock
LST	Landing Ship, Tank

Shore Stations:

Call	Location
NNN0EBC	Stateside
NNN0KRQ	Cincinatti, Ohio
NNN0MQU	Marine Base, Virginia
NNN0NAV	MARS HQ, Washington, DC
NNN0NHA	Norfolk, Virginia
NNN0NIK	Mayport, Florida
NNN0PPE	Vienna, Virginia
NNN0PVU	Norfolk, Virginia
NNN0QCA	West Winfield, New York
NNN0QGR	Middletown, Rhode Island
NNN0QLP	Middletown, Rhode Island
NNN0RAM	Portsmouth, Virginia
NNN0USN	Dahlgren, Virginia
NNN0UXK	Beaver Dam, Wisconsin
NNN0VKC	Jacksonville, Florida
NNN0WHT	Springfield, Massachusettes
NNN0XPQ	Fort Lauderdale, Florida
NNN0YPD	Lutherville, Maryland
NNN0ZEC	South Carolina
NNN0ZEJ	Virginia Beach, Virginia
NNN0ZIW	Norfolk, Virginia

NNNOZMH Stateside
NNNOZTI Virginia Beach, Virginia

Although both ship and shore station call-signs are prefixed by NNNO, it is usually excluded from any communications.

Within the amateur bands, N6ZO Maritime Mobile Region 3 and K3FT Maritime Mobile Region 1 on board the hospital ship USS Comfort in the Gulf is currently passing phone patch traffic to several amateur stations in the US. The net meets almost nightly around 14346.0 kHz USB, and remains active for several hours. Operator chit-chat is reminiscent of the humour from the TV programme MASH! Callsigns in the net include:

K4EWS	
K4QJN	
WA1GUZ	
WA1PMC	Palm Harbour, Florida
W2PCZ	

ROYAL NAVY

Unfortunately for radio monitors, the Royal Navy are a bit shy when it comes to disclosing their tactical communications and rely almost totally on the SKYNET satellites to relay information to and from the fleet coordination centre at Northwood in North London. Frequencies for these satellites are now known to be up in the 7 GHz range, and are therefore out of the range of most listeners. For those who do have access, the downlink frequencies are:

2 MHz channel	: 7257.3-7259.9 GHz
20 MHz channel	: 7266.4-7286.4 GHz
Beacon	: 7299.1 GHz
Mobile terminal	: around 250 MHz
Skynet 2	: around 370-400 MHz

Some Royal Navy ships have been heard sending traffic to the UK's maritime radio station at Portishead, so it may be worth while keeping a watch on these frequencies.

Portishead Radio, GKU, United Kingdom

Mode: USB

Ship	Shore	Call
8198.1	8722.0	GKT 42
8250.8	8774.7	GKU 49
8272.5	8796.4	GKV 46
8288.0	8811.9	GKW 41
12330.0	13100.8	GKT 51
12345.5	13116.3	GKT 56
12401.3	13172.1	GKV 54
12426.1	13196.9	GKW 52
16491.0	17263.9	GKU 61
16503.4	17276.3	GKU 65
16512.7	17285.6	GKU 68
16571.6	17344.5	GKW 67

The Royal Navy does however operate several HF networks. The most widely known being that of the Morse station in Gibraltar. I have monitored this station, but have only found the usual CW marker, with no apparent traffic. The frequencies so far monitored as active include:

8270.0 kHz   12824.2   16987.2

Royal Navy ships presently in the Gulf Area are:

Ships Name	Callsign	Type	Hull No.
RFA Argus	GDSA	Hospital Ship	
HMS Atherstone	GBPJ	Minehunter	M36
HMS Battleaxe	GVEW	Frigate	F89
HMS Brazen	GZIT	Frigate	F91
HMS Cardiff	GQIT	Destroyer	D109
HMS Cattistock	GYHK	Minehunter	M31
HMS Gloucester	GBBF	Destroyer	D96
HMS Hurworth	GBPL	Minehunter	M39
HMS Jupiter	GSYA	Frigate	F60
HMS London	GDBV	Frigate	F95
RFA Orangeleaf	GURK	Supply Ship	A110
HMS Sir Bedivere	GSRE	Landing Ship	L3004
HMS Sir Galahad	GABN	Landing Ship	
HMS Sir Tristram	GVMA	Landing Ship	L3505
HMS York	GBBB	Destroyer	D98

US ARMY

In the first edition of this book, US Army MARS stations had not yet established themselves. Now they have and are active on a nightly basis. At the present time, net security is relatively low with QSY information being passed in kilohertz, but the Army is now introducing a designator system to all units. References to the 'Sandbox' may be heard on these nets. This is the US Army MARS code-name for Saudi Arabia.

US Army MARS activity can be found on the following frequencies:

Mode: USB

Frequency	Designator
14633.5 kHz	C04
14666.0	C05
20811.0	E04
20940.0	E06

Additional frequencies include:

13714 kHz	14512
13741	14579
13992	14580
13993	14633.5
14402	14666
14404	20811
14485	20940
14487	20991
14510	20993
14511	20994

Callsigns on these frequencies include:

Call	Location
AAR2NP	New York
AAR3NAA	Stateside
AAR4LL	Stateside

AAR4USC	Ft. Bragg, N. Carolina
AAR4USD	Ft. Jackson, S. Carolina
AAR4USW	Military Terminal, New Jersey
AAR4USZ	Ft. McClellan, Alabama
AAT9AJ	Southern California
AAV4AN	Augusta, Georgia
AAV4RN	Stateside
AAV4RR	Stateside
AAV4TN	Florida
AAV6GG	Stateside
AAV6KG	Oklahoma
AEM1GST	Europe
AEM1PV	Europe
AEM1QF	Berlin, Germany
AEM3UA	Saudi Arabia
AEM3UC	Saudi Arabia
AEM3UF	Saudi Arabia
AEM3UH	Saudi Arabia
AEM3UO	Saudi Arabia
AEM3UP	Saudi Arabia
AEM3UR	Saudi Arabia
AEM3UQ	Saudi Arabia
AEM3UT	Saudi Arabia
AEM3UU	Saudi Arabia

Note that any station which has a callsign beginning with AEM3 is a Gulf unit based in the "Sandbox." The above list only contains those stations which have previously been monitored, but the series does run from AEM3UA to AEM3UZ allowing for 26 individual stations.

Station AEM3UA was previously known as AAR0USA and was the first US Army MARS station to broadcast from Saudi Arabia. It was initially being operated by the stateside US Navy MARS station NNN0RWO in Myrtle Beach, South Carolina, but since other stateside Army MARS units have become active, RWO has severed his daily 2300hrs schedule, and is now running traffic for units linked to Myrtle Beach AFB.

GULF AVIATION

The Gulf region, Air Route Area MID, centred on northern Iran, is one of the busiest aviation lanes in the world. Aircraft in the area report to six principal HF Air Traffic Control (ATC) centres located in the general Gulf area.

Gulf Aero traffic: Mode: USB

3095 kHz	Saudi Airports	
3404 kHz	Baghdad	Beirut
	Basrah	Damascus
3453 kHz	Saudi Airports	
5603 kHz	Amman	Beirut
	Baghdad	Damascus
	Basrah	
5658 kHz	Bombay	Manama
	Jeddah	Muscat
	Kabul	New Delhi
	Karachi	Tehran
	Kuwait	
5667 kHz	Amman	Jeddah
	Baghdad	Kuwait
	Basrah	Manama
	Damascus	
8918 kHz	Amman	Jeddah
	Baghdad	Kuwait
	Basrah	Manama
	Damascus	
8967 kHz	Saudi Airports (0400-1900hrs UTC)	
8990 kHz	Jeddah	
	Riyadh	

10018 kHz	Kuwait Muscat	
13288 kHz	Kuwait Manama	
13312 kHz	Amman Basrah Beirut	Jeddah Kuwait

Many of the worlds larger airlines maintain an HF radio network for the benefit of their pilots on long distance flights. It was recently reported that the ill-fated BA 147 flight which was stranded in Kuwait at the beginning of the invasion, attempted to contact another BA flight in the area on a "discrete company channel!"

British Airways, London — Callsign: Speedbird London — Mode : USB

3481 kHz	10072
3497	13333
5535	17922
6568	21946
8921	

Gulf Air, Manama, Bahrain — Callsign: Falcon — Mode : USB

5538 kHz	17931
8811	21943
13339	

Royal Jordanian, Amman — Callsign: Alia Base — Mode : USB

9003 kHz

Saudia, Jeddah — Callsign: Jeddah — Mode : USB

4710 kHz	13342
5544	17925
8921	21994
8927	23307
13339	

For those airlines which do not themselves maintain long distance HF radio communications with their aircraft, it is worth checking out some of the commercial aviation stations, such as Berne, Portishead and Stockholm Radios. For it is with these stations that stranded aircraft will try and contact their home airports.

Berne Radio, Switzerland — Mode: USB

4654 kHz	18023
6643	20870
8936	21988
13324	23285
15046	25500
15835	

Portishead Radio, United Kingdom — Mode: USB

4492 kHz	8185	13865	20065
4807	8960	14890	21765
5610	10291	16370	
6634	12168	17405	
8170	12168	18210	

Stockholm Radio, Sweden — Mode: USB

5290 kHz	11345	15685	23210
5541	13342	17916	
10165	13576	17982	
10575	13942	20770	
11222	15021	21997	

Recently, an American military transport aircraft was monitored working Stockholm Radio. The aircraft, call-sign MAC 819, had a communications break down, and the conversation eventually deteriorated to:

"Are you happy about going to Mildenhall? Key your mic twice for Yes, or once for No!"

"Click, click..."

SAM flights have also been monitored working Stockholm Radio from the Gulf area!

GULF SHIPPING

During the eight year Iran-Iraq Gulf war, the shipping lanes of the Persian Gulf were one of the most hostile areas in the conflict. On these same water-ways is shipped well over half the worlds oil imports, and with the present UN security regulations, both British and American warships can board any tanker that is suspected of breaking the Iraq trade embargo.

Coastal stations in the heart of the conflict are Safat Radio in Kuwait, Basrah Control in Iraq and Aqaba Radio in Jordan. None of these stations have been monitored since the beginning of the conflict, which would apparently point to UN trade embargos taking effect.

Basrah Control, YIR, Iraq

Mode: USB		Mode: CW	
Ship	Shore		
4069.2 kHz	4363.6 kHz	4220.0 kHz	16880.0
6200.0	6506.4	6330.0	16906.0
8238.4	8762.3	8440.0	22338.0
12336.2	13107.0	12660.0	
16537.5	17310.4		
22037.2	22633.2		

Safat Radio, 9KK, Kuwait

Mode: USB		Mode: CW	
Ship	Shore		
4118.8 kHz	4413.2 kHz	6381.0 kHz	12925.0
4118.8	6509.5	8525.0	16995.0
8213.6	8737.5	12895.0	22504.0
12410.6	13181.4		
16515.8	17288.7		
22009.3	22605.3		

The Jordanian Red Sea port of Aqaba has developed a crisis all of its own. It is the only port that could accept trade to be transported to Iraq, and is being watched 24 hours a day by US Navy vessels. This embargo, in force for well over two months, now seems to be taking effect.

Aqaba Radio, JYO, Jordan

Mode: USB		Mode: CW	
Ship	Shore		
4121.9 kHz	4416.3 kHz	4326.5 kHz	8528.0
4137.4	4431.8	6390.0	12725.0
8204.3	8728.2	6479.0	
12388.9	13159.7		

Other coastal stations in the Gulf are carrying on work as normal. These include:

Bahrain Radio, A9M, Bahrain
Mode: CW

12709.0 kHz
17169.0
17175.2
22322.0

Damman Radio, HZG, Saudi Arabia
Mode: CW

8484.5 kHz
12658.2
16860.8
22338.2

Both of these stations can be heard with their CW markers, DE A9M and DE HZG respectively. Traffic lists are also sent, and it may be worth listening out for Kuwaiti ships in these lists, since they are now without a country or home port.

Another set of frequencies worth keeping in mind are the international emergency and distress channels:

2182.0 kHz	8257.0
4125.0	12429.2
6218.6	16522.0

INTELLIGENCE COMMUNICATIONS

CENTRAL INTELLIGENCE AGENCY

Ever since the commencement of hostilities in the Gulf region, US President and former Director of CIA, George Bush, has been constantly instructing the Central Intelligence Agency in Langley, Virginia, to assess the feasibility of plans to overthrow the Baghdad regime. Rumours have also been circulated that the CIA are planning attempts to assassinate Iraq leader, Saddam Hussein.

The validity of such rumours have been somewhat reinforced by a general increase in traffic from the CIA's European English language number stations known as "Cynthia from Langley." Since the beginning of August, 1990, scheduled traffic continued to increase to a state where messages were being aired on an almost hourly basis. Towards the end of September, traffic began to steadily decrease returning to its present pre-crisis level.

Presently, Cynthia is residing on the following frequencies:

Mode: USB

9096 kHz	14703
9147	14752
9844	14811
11034	15938
11514	16377
12230	18480
13415	20115
14355	20524
14448	

The American intelligence community, in particular, the CIA and NSA, analyse literally hundreds of thousands of pieces of information every day, basing future policies and defence decisions on what they find. Information from the Gulf is presently being collected by a family of covert surveillance satellites, some of which are believed to have been launched by the Space Shuttle. The satellites, known as KH-11, Keyhole, Rhyolite and Magnum, orbiting between 500 and 23,500 miles above the earth, listen in on everything from microwave telephone calls to signatures from ground radar stations.

MOSSAD (Selected Traffic 0700–2330hrs UTC)

The largest Intelligence agency in the Gulf region is undoubtedly the Israeli Mossad. Like the CIA, it too maintains an HF network.

Broadcast markers: Mode: USB

Each broadcast usually lasts 5 minutes. The best marker to monitor is CIO2. Any deviation from its schedule or format usually means that something important is going to happen. This beacon has also been linked to Israeli Army manoeuvres within Israel.

CIO2: Broadcasts 45 minutes after the hour.
0700–1400hrs UTC on 13921, 17966 and 23195 kHz
1500–2100hrs UTC on 10125, 13921 and 17966 kHz
2200–2300hrs UTC on 6745, 10125 and 13921 kHz

KPA2: Broadcasts 15 minutes after the hour.
0700–1800hrs UTC on 10820 kHz
1900–2300hrs UTC on 3270, 4780 and 7445 kHz

MIW2: Broadcasts 15 minutes after the hour.
0700–1500hrs UTC on 12747, 17170 and 20740 kHz
1600–2300hrs UTC on 8641, 12747 and 17170 kHz

SYN2: Broadcasts 30 minutes after the hour.
0700–1500hrs UTC on 8465 and 12950 kHz
1600–2300hrs UTC on 5629 and 8465 kHz

VLB2: Broadcasts 45 minutes after the hour.
0700–2345hrs UTC on 12950, 14750 and 18178 kHz

Traffic Broadcasts: Mode: USB

These broadcasts consist of five letter crypto groups and are broadcast on the hour and half hour. These transmissions do not seem to reflect current events in the same way as the markers do.

ART: 1800-2300hrs UTC on 3417 and 5437 kHz

EZI: 0900-1100hrs UTC on 11565 kHz
1130-1200hrs UTC on 13533 and 15980 kHz
1700hrs UTC on 17410 and 19715 kHz
1730-1800 & 2000hrs UTC on 9130 and 11565 kHz
1900, 2200 & 2230hrs UTC on 6840 and 9130 kHz
2100hrs UTC on 11565 kHz

JSR: 1800-2330hrs UTC on 2270 and 5091 kHz

PCD: 1930-2330hrs UTC on 3150 and 4270 kHz

ULX: 1900-2330hrs UTC on 2743 and 4880 kHz

YHF: 1800-1830hrs UTC on 5820 and 10648 kHz
1930-2000 & 2100-2330hrs UTC on 3840 and 4560 kHz
2030hrs UTC on 3840 and 7918 kHz

PALESTINE LIBERATION ORGANISATION

Although not exactly an intelligence agency, the PLO offers an interesting target in the Middle East. PLO offices in Aden and Damascus are listed as transmitting in USB on 14390 kHz, but this is not the only frequency.

In recent weeks, transmissions have been heard on:

Mode: USB or LSB

14380 kHz	14393
14388	14395
14390	14399

Transmissions are in Arabic and can be heard in the early morning and evening. Traffic consists of radio checks, operator chit-chat, and plain text messages characterised by phrases and sentences repeated twice. The closing lines usually contain the name of PLO chairman Yasser Arafat.

A second unidentified Arabic net has recently become active. The net operates at night on 10100.0 kHz in LSB or USB, and consists of operator chit-chat and the odd song during which all operators in the net join in!

MONITORING NEWS FROM THE GULF

This is not a true guide to all broadcasting stations in the Gulf area, but only reflects some of the more important English transmissions. For a full schedule, I recommend that you consult a recognised broadcasting guide.

English News schedules:

Radio Baghdad, Iraq

2000-2300hrs UTC on 13660 kHz

This station is still furnishing the bulk on anti-American propaganda from the Gulf area. Nonetheless, this station has a habit of giving some very accurate reports.

Radio Cairo, Egypt

2115-2245hrs UTC on 9900 kHz

Cairo initially took the stance that diplomatic talks were the key to solving the crisis, but since they sent 4000 ground troops, their views have altered somewhat. Signals are generally of a very poor quality, which does not produce easy listening.

Radio Damascus, Syria

2010-2105hrs UTC on 12085 and 15095 kHz (Europe)
2110-2205hrs UTC on 12085 and 15095 kHz (Asia)

The arch-enemy of Iraq is producing propaganda to that effect, although news is very accurate. This station broadcasts some of the best music from the Gulf area. Main news is at the beginning of the broadcast with a short summary at the end.

Kol Israel, Jerusalem, Israel

2230-2300hrs UTC on 11585, 11605, 15640 and 17575 kHz
0100-0130hrs UTC on 11585 and 11605 kHz

As piggy in the middle, Israel is riding a precarious road, firmly believing that Iraq will change its sights from Kuwait to Israel at any time. Therefore news is as always, pro-American, although on Israels own admission, it wishes to keep a very low profile.

UAE Radio, Dubai, UAE

1030-1100hrs UTC on 15320, 15435, 21605 and 21675 kHz
1330-1400hrs UTC on 15320, 15435, 21605 and 21675 kHz
1600-1645hrs UTC on 15320, 15435, 21605 and 21675 kHz

This is undoubtedly one of the best sources of news from the Gulf and should be your number one choice. Although all four frequencies are announced at 1030hrs, only 21675 kHz, which is shared with Radio Moscow, seems to get through.

Voice of Islamic Republic of Iran, Tehran, Iran

1930-2030hrs UTC on 9022 kHz

After the eight year war with Iraq, Iran is very suspicious of the hand of friendship that Saddam Hussein is offering, and their news reports reflect this view.

Voice of Turkey, Ankara, Turkey

2100-2150hrs UTC on 9725 kHz
2300-2350hrs UTC on 9445 and 9685 kHz

The Voice of Turkey presents a very accurate and impartial news cast about the Gulf, so its worth tuning to this station to find out what's going on.

Voice of UAE, Abu Dhabi, UAE

2200-2400hrs UTC on 9600, 11965 and 13605 kHz

This station is now concentrating on Arabic transmissions.

B.S.K.S.A., Riyadh, Saudi Arabia

1600-2100hrs UTC on 9705 and 9720 kHz

Radio Jordan, Amman, Jordan

1320-1630hrs UTC on 9560 kHz

English broadcasts are not aired from Qatar, Oman or Yemen, while Bahrain depended on Kuwaiti HF transmitter facilities.

All India Radio, New Delhi

1845-1945hrs UTC on 7412 and 11620 kHz
2045-2230hrs UTC on 7412 and 11620 kHz

British Forces Broadcasting Service

1330-1400hrs UTC on 17695 and 21735 kHz

The following three stations, namely BBC, RMWS and VOA all have rather complicated schedules which would each require several pages to list. So in order to conserve space, I have simply listed the frequencies that each station can be heard on throughout the day.

BBC, London

3955 kHz	5975	6175	6180	7325	9410	9750
9760	9915	11850	12095	15070	15260	15400

And for people living in Europe, BBC Radio 4 on 198 kHz relays the World Service between 0045 and 0545hrs UTC.

Radio Moscow World Service, Moscow

5980 kHz	7175	7350	9450	9520	9610	11520
11730	11745	11775	11920	11930	11985	12030
15140	15210	15220	15470	15540	15550	15560
17570	17580	17655	17730	17775	17810	17830
21585	21630	21715	21725	21755	21785	25780

Voice of America, Washington DC

5995 kHz	6040	7200	7325	9700	15160	15185
15205	15225	15290	15410	15580		

CHANNEL DESIGNATORS

This list of channel designators contains all the most important and commonly used channels. Many have already been used in association with the Gulf, while other will most definately be used in an escalating conflict. The list is far from complete and only reflects frequencies where activity is likely to be heard that designators are available for.

Designator	kHz	Mode	User
ACA	11407.0	USB	USAF MARS
ACB	13927.0	USB	USAF MARS
ACE	14390.5	USB	USAF MARS
ACF	14606.0	USB	USAF MARS
ACG	20188.5	USB	USAF MARS
C04	14633.5	USB	US Army MARS
C06	14666.0	USB	US Army MARS
A	11121.0	USB	USAF MARS
B	11407.0	USB	USAF MARS
D	13927.0	USB	USAF MARS
E	14382.0	USB	USAF MARS
E04	20811.0	USB	US Army MARS
E06	20940.0	USB	US Army MARS
F	14405.0	USB	USAF MARS
F24	6993.0	U/LSB	Andrews AFB
F40	11056.0	U/LSB	Andrews AFB
F43	11466.0	U/LSB	Andrews AFB
F95	9043.0	U/LSB	Andrews AFB
F96	11407.0	U/LSB	Andrews AFB
F99	13878.0	U/LSB	Andrews AFB
F114	5800.0	U/LSB	Andrews AFB
F115	6812.0	U/LSB	Andrews AFB
F140	18175.0	U/LSB	Andrews AFB
F151	6756.0	U/LSB	Andrews AFB
F152	11226.0	U/LSB	Andrews AFB
F158	15015.0	U/LSB	Andrews AFB
F166	23035.0	U/LSB	Andrews AFB
F182	6993.0	U/LSB	Andrews AFB
F195	8162.0	U/LSB	Andrews AFB
F200	9320.0	U/LSB	Andrews AFB
F202	9414.0	U/LSB	Andrews AFB
F204	10583.0	U/LSB	Andrews AFB

F212	11052.0	U/LSB	Andrews AFB
F213	11055.0	U/LSB	Andrews AFB
F214	11058.0	U/LSB	Andrews AFB
F220	11488.0	U/LSB	Andrews AFB
F290	6817.0	U/LSB	Andrews AFB
F292	7997.0	U/LSB	Andrews AFB
F299	9320.0	U/LSB	Andrews AFB
F301	10112.0	U/LSB	Andrews AFB
F304	11056.0	U/LSB	Andrews AFB
F328	18320.0	U/LSB	Andrews AFB
F350	5820.0	U/LSB	Andrews AFB
F352	6993.0	U/LSB	Andrews AFB
F361	9120.0	U/LSB	Andrews AFB
F366	10427.0	U/LSB	Andrews AFB
F369	11056.0	U/LSB	Andrews AFB
F370	11466.0	U/LSB	Andrews AFB
F377	13823.0	U/LSB	Andrews AFB
F388	18320.0	U/LSB	Andrews AFB
F403	6993.0	U/LSB	Andrews AFB
F415	11466.0	U/LSB	Andrews AFB
F416	11484.0	U/LSB	Andrews AFB
F435	18532.0	U/LSB	Andrews AFB
F444	5760.0	U/LSB	Andrews AFB
F455	11118.0	U/LSB	Andrews AFB
F461	18320.0	U/LSB	Andrews AFB
F468	6918.0	U/LSB	Andrews AFB
F469	7858.0	U/LSB	Andrews AFB
F474	13205.0	U/LSB	Andrews AFB
F479	23385.0	U/LSB	Andrews AFB
F484	6830.0	U/LSB	Andrews AFB
F486	8040.0	U/LSB	Andrews AFB
F487	9120.0	U/LSB	Andrews AFB
F511	15680.0	U/LSB	Andrews AFB
F528	5710.0	U/LSB	Andrews AFB
F529	6683.0	U/LSB	Andrews AFB
F530	6716.0	U/LSB	Andrews AFB
F531	6730.0	U/LSB	Andrews AFB
F535	11226.0	U/LSB	Andrews AFB
F537	13217.0	U/LSB	Andrews AFB
F541	17972.0	U/LSB	Andrews AFB
G	14607.5	USB	USAF MARS
J	17487.0	USB	USAF MARS
O	20737.0	USB	USAF MARS
P	20992.5	USB	USAF MARS

P-383	15044.0	USB	USAF SAC
S-302	3113.0	USB	USAF SAC
S-303	3295.0	USB	USAF SAC
S-304	4495.0	USB	USAF SAC
S-309	9057.0	USB	USAF SAC
S-310	11220.0	USB	USAF SAC
S-311	11494.0	USB	USAF SAC
S-312	13211.0	USB	USAF SAC
S-391	6761.0	USB	USAF SAC
S-392	9027.0	USB	USAF SAC
S-393	11243.0	USB	USAF SAC
S-394	13241.0	USB	USAF SAC
S-395	17975.0	USB	USAF SAC
W-101	5700.0	USB	USAF SAC
W-103	6757.0	USB	USAF SAC
W-104	7475.0	USB	USAF SAC
W-105	7831.0	USB	USAF SAC
W-108	12070.0	USB	USAF SAC
W-109	13247.0	USB	USAF SAC
W-111	17972.0	USB	USAF SAC
W-116	20167.0	USB	USAF SAC
X-904	9017.0	USB	USAF SAC
X-905	11226.0	USB	USAF SAC

The only list of UHF designators appertains to SAM and Air Force One/Two flights. The user being Andrews AFB, with the mode being either WFM, USB, LSB or NBAM.

Designator	Ground	Air
RF1	326.000 MHz	382.350 MHz
RF2	246.950	305.550
RF3	345.500	336.800
RF4	322.750	366.000
RF5	390.000	397.050
RF6	382.350	326.000
RF7	305.550	246.950
RF8	336.800	345.500
RF9	366.000	322.750
RF10	397.050	390.000

CALLSIGN LIST

Callsign	Location
US Army MARS:	
AAR2NP	New York
AAR3NAA	Stateside
AAR4LL	Stateside
AAR4USC	Ft. Bragg, N. Carolina
AAR4USD	Ft. Jackson, S. Carolina
AAR4USW	Military Terminal, New Jersey
AAR4USZ	Ft. McClellan, Alabama
AAT9AJ	Southern California
AAV4AN	Augusta, Georgia
AAV4RN	Stateside
AAV4RR	Stateside
AAV4TN	Florida
AAV6GG	Stateside
AAV6KG	Oklahoma
AEM1GST	Europe
AEM1PV	Europe
AEM1QF	Berlin, Germany
AEM3UA	Saudi Arabia
AEM3UC	Saudi Arabia
AEM3UF	Saudi Arabia
AEM3UH	Saudi Arabia
AEM3UO	Saudi Arabia
AEM3UP	Saudi Arabia
AEM3UR	Saudi Arabia
AEM3UQ	Saudi Arabia
AEM3UT	Saudi Arabia
AEM3UU	Saudi Arabia
US Air Force MARS:	
AFA1MH	Stateside
AFA1YV	Stateside
AFA2DE	Stateside
AFA2KD	Stateside
AFA2SP	Stateside
AFA2XD	Stateside
AFA3BZ	Stateside
AFA4KE	Kelly AFB, Texas

AFA7BQ	Europe
AFA7DW	Europe
AFA7MJ	Ramstein AB, Germany
AFA7NL	Europe
AFA7TJ	Europe
AFB2DE	Ft Lauderdale, Florida
AFB2RF	Stateside
AFF2SP	Miami, Florida
AFQ375	Saudi Arabia
AFQ376	Saudi Arabia
AFQ379	Saudi Arabia
AFQ472	Saudi Arabia
AFT7MP	Europe
AGA1MC	MacGuire AFB, New Jersey
AGA2MB	Myrtle Beach AFB, S. Carolina
AGA2PA	Patrick AFB, Florida
AGA2PO	Pope AFB, North Carolina
AGA2PX	Deployed Unit in Saudi Arabia
AGA3HQ	Scott AFB, Illinois
AGA7BI	Bitburg AB, Germany
AGA7DW	Europe
AGA7IR	Iraklion AB, Greece
AGA7LI	Lindsey AB, Germany
AGA7MJ	Ramstien, Germany
AGA7ML	Europe
AGA7RM	Rhein Main AB, Germany
AIR	USAF MARS HQ, Andrews AFB
CUW20	Net Coordinator, Lajes AB, Azores
ENN1183	Saudi Arabia

US Navy MARS:

NNN0CAI	USS Antietam
NNN0CBD	USS Thomas S. Gates
NNN0CJN	USS Reid
NNN0CMU	USS Seattle
NNN0CNE	USS England
NNN0COA	USS Raleigh
NNN0COE	USS Independence
NNN0COO	USS Tattnal
NNN0COQ	USS Pensacola
NNN0COU	USS Saratoga
NNN0CSB	USS Spartanburg County
NNN0CTQ	USS O'Brien

NNNOCTN	USS Barbey
NNNOCUX	USS Nassau
NNNOCVA	USS Belknap
NNNOCVG	USS Dwight D. Eisenhower
NNNOCVU	USS Saginaw
NNNOCWI	USS Wisconsin
NNNOCWO	USS Flint
NNNOCXB	USS John F. Kennedy
NNNOCXN	USS Portland
NNNOEBC	Stateside
NNNOKRQ	Cincinatti, Ohio
NNNOMQU	Marine Base, Virginia
NNNONAV	MARS HQ, Washington, DC
NNNONHA	Norfolk, Virginia
NNNONIK	Mayport, Florida
NNNONXK	USS Iwo Jima
NNNONZR	USS La Salle
NNNOPPE	Vienna, Virginia
NNNOPVU	Norfolk, Virginia
NNNOQCA	West Winfield, New York
NNNOQGR	Middletown, Rhode Island
NNNOQLP	Middletown, Rhode Island
NNNORAM	Portsmouth, Virginia
NNNOUSN	Dahlgren, Virginia
NNNOUXK	Beaver Dam, Wisconsin
NNNOVKC	Jacksonville, Florida
NNNOWHT	Springfield, Massachusettes
NNNOXPQ	Fort Lauderdale, Florida
NNNOYPD	Lutherville, Maryland
NNNOZEC	South Carolina
NNNOZEJ	Virginia Beach, Virginia
NNNOZIW	Norfolk, Virginia
NNNOZMH	Stateside
NNNOZTI	Virginia Beach, Virginia

ARCHITECT	RAF Upavon Ground Control, United Kingdom
CROUGHTON	USAF GCCS, RAF Croughton, United Kingdom
FORMAT	MAC HQ USA, Scott AFB, Illinois
INCIRLIK	USAF GCCS, Incirlik AFB, Turkey
PHANTOM	MAC HQ Europe, Ramstein AFB, Germany
WATERLOO	Unknown MAC station

FREQUENCY LIST

This list contains all the frequencies that have been, or are likely to be active with Gulf traffic.

Freq	Mode	Call	User
2512.5	USB	JYO	Aqaba Radio, Jordan
2612.6	USB	JYO	Aqaba Radio, Jordan
2812.0	USB		International Distress Chan.
3067.0	USB	Croughton	USAF GCCS
3081.0	USB	Lajes	USAF GCCS
3095.0	USB		Gulf ATC
			Royal Saudi Air Force
3404.0	USB		Gulf ATC
3453.0	USB		Gulf ATC
3481.0	USB	Speedbird	British Airways Company Chan.
3497.0	USB	Speedbird	British Airways Company Chan.
4069.2	USB	YIR	Basrah Control, Iraq
4121.9	USB	JYO	Aqaba Radio, Jordan
4125.0	USB		International Distress Chan.
4137.4	USB	JYO	Aqaba Radio, Jordan
4200.0	CW	YIR	Basrah Control, Iraq
4326.5	CW	JYO	Aqaba Radio, Jordan
4363.6	USB	YIR	Basrah Control, Iraq
4416.3	USB	JYO	Aqaba Radio, Jordan
4431.8	USB	JYO	Aqaba Radio, Jordan
4492.0	USB	Portishead	Portishead Aero Radio
4654.0	USB	Berne	Berne Aero Radio
4702.0	USB		USN Tactical Channel
4710.0	USB	Jeddah	Saudia Company Channel
4711.0	USB		USN Tactical Channel
4725.0	USB		Strategic Air Command
4730.0	USB	Cyprus	RAF Akrotiri
4742.0	USB	Architect	RAF Upavon
4746.0	USB	Lajes	USAF GCCS
4807.0	USB	Portishead	Portishead Aero Radio
5290.0	USB	Stockholm	Stockholm Aero Radio
5535.0	USB	Speedbird	British Airways Company Chan.
5538.0	USB	Falcon	Gulf Air Company Channel
5541.0	USB	Stockholm	Stockholm Aero Radio
5544.0	USB	Jeddah	Saudia Company Channel

5603.0	USB		Gulf ATC
5610.0	USB	Portishead	Portishead Aero Radio
5658.0	USB		Gulf ATC
5667.0	USB		Gulf ATC
5683.0	USB		Kuwaiti Air Force
5703.0	USB	Croughton	USAF GCCS
5710.0	L/USB	Andrews	USAF Mystic Star Channel
5713.0	USB	Architect	RAF Upavon
5760.0	L/USB	Andrews	USAF Mystic Star Channel
5800.0	USB		Strategic Air Command
	L/USB	Andrews	USAF Mystic Star Channel
5820.0	L/USB	Andrews	USAF Mystic Star Channel
6200.0	USB	YIR	Basrah Control, Iraq
6218.6	USB		International Distress Chan.
6330.0	CW	YIR	Basrah Control, Iraq
6390.0	CW	JYO	Aqaba Radio, Jordan
6479.0	CW	JYO	Aqaba Radio, Jordan
6506.4	USB	YIR	Basrah Control, Iraq
6568.0	USB	Speedbird	British Airways Company Chan.
6643.0	USB	Berne	Berne Aero Radio
6683.0	L/USB	Andrews	USAF Mystic Star Channel
6716.0	L/USB	Andrews	USAF Mystic Star Channel
6720.0	USB		USN HICOM Atlantic Ocean
6730.0	L/USB	Andrews	USAF Mystic Star Channel
6738.0	USB	Architect	RAF Upavon
		Incirlik	USAF GCCS
		Lajes	USAF GCCS
6745.0	USB	CIO2	MOSSAD
6750.0	USB	Croughton	USAF GCCS
6756.0	L/USB	Andrews	USAF Mystic Star Channel
6757.0	USB		Strategic Air Command
6761.0	USB		Strategic Air Command
6770.0	USB		USN Tactical Channel
6790.0	L/USB	Andrews	USAF Mystic Star Channel
6804.0	USB		USN Tactical Channel
6812.0	L/USB	Andrews	USAF Mystic Star Channel
6817.0	L/USB	Andrews	USAF Mystic Star Channel
6830.0	L/USB	Andrews	USAF Mystic Star Channel
6918.0	L/USB	Andrews	USAF Mystic Star Channel
6927.0	L/USB	Andrews	USAF Mystic Star Channel
6993.0	L/USB	Andrews	USAF Mystic Star Channel
6996.0	USB		USAF MARS
7475.0	USB		Strategic Air Command
7813.0	L/USB	Andrews	USAF Mystic Star Channel

7831.0	USB		Strategic Air Command
7858.0	L/USB	Andrews	USAF Mystic Star Channel
7997.0	L/USB	Andrews	USAF Mystic Star Channel
8040.0	L/USB	Andrews	USAF Mystic Star Channel
8162.0	L/USB	Andrews	USAF Mystic Star Channel
8170.0	USB	Portishead	Portishead Aero Radio
8185.0	USB	Portishead	Portishead Aero Radio
8204.3	USB	JYO	Aqaba Radio, Jordan
8238.4	USB	YIR	Basrah Control, Iraq
8257.0	USB		International Distress Chan.
8440.0	CW	YIR	Basrah Control, Iraq
8465.0	USB	SYN2	MOSSAD
8484.5	CW	HZG	Damman Radio, Saudi Arabia
8528.0	CW	JYO	Aqaba Radio, Jordan
8722.0	USB	Portishead	Portishead Maritime Radio
8728.2	USB	JYO	Aqaba Radio, Jordan
8762.3	USB	YIR	Basrah Control, Iraq
8774.7	USB	Portishead	Portishead Maritime Radio
8796.4	USB	Portishead	Portishead Maritime Radio
8811.0	USB	Falcon	Gulf Air Company Channel
8811.9	USB	Portishead	Portishead Maritime Radio
8918.0	USB		Gulf ATC
8921.0	USB	Jeddah	Saudia Company Channel
		Speedbird	British Airways Company Chan.
8927.0	USB	Jeddah	Saudia Company Channel
8936.0	USB	Berne	Berne Aero Radio
8967.0	USB	Lajes	USAF GCCS
			Royal Saudi Air Force
			Gulf ATC
8970.0	USB		Kuwaiti Air Force
8990.0	USB		Gulf ATC
9003.0	USB	Alia Base	Royal Jordanian Company
9011.0	USB	Croughton	USAF GCCS
9017.0	USB		Strategic Air Command
	L/USB	Andrews	USAF Mystic Star Channel
9018.0	L/USB	Andrews	USAF Mystic Star Channel
9022.0	AM		Voice of Islamic Rep. of Iran
9027.0	USB		Strategic Air Command
9032.0	USB	Architect	RAF Upavon
		Cyprus	RAF Akrotiri
9043.0	L/USB	Andrews	USAF Mystic Star Channel
9057.0	USB		Strategic Air Command
	L/USB	Andrews	USAF Mystic Star Channel
9096.0	USB		Central Intelligence Agency

9120.0	L/USB	Andrews	USAF Mystic Star Channel
9147.0	USB		Central Intelligence Agency
9270.0	L/USB	Andrews	USAF Mystic Star Channel
9320.0	L/USB	Andrews	USAF Mystic Star Channel
9414.0	L/USB	Andrews	USAF Mystic Star Channel
9435.0	AM		Kol Israel, Jerusalem
9445.0	AM		Voice of Turkey
9560.0	AM		Radio Jordan
9600.0	AM		Voice of UAE, Abu Dhabi
9685.0	AM		Voice of Turkey
9725.0	AM		Voice of Turkey
9844.0	USB		Central Intelligence Agency
9900.0	AM		Radio Cairo
9950.0	AM		Radio Damascus
10018.0	USB		Gulf ATC
10072.0	USB	Speedbird	British Airways Company Chan.
10112.0	L/USB	Andrews	USAF Mystic Star Channel
10125.0	USB	CIO2	MOSSAD
10165.0	USB	Stockholm	Stockholm Aero Radio
10291.0	USB	Portishead	Portishead Aero Radio
10427.0	L/USB	Andrews	USAF Mystic Star Channel
10575.0	USB	Stockholm	Stockholm Aero Radio
10583.0	L/USB	Andrews	USAF Mystic Star Channel
10820.0	USB	KPA2	MOSSAD
11034.0	USB		Central Intelligence Agency
11035.0	L/USB	Andrews	USAF Mystic Star Channel
11052.0	L/USB	Andrews	USAF Mystic Star Channel
11055.0	L/USB	Andrews	USAF Mystic Star Channel
11056.0	L/USB	Andrews	USAF Mystic Star Channel
11058.0	L/USB	Andrews	USAF Mystic Star Channel
11118.0	L/USB	Andrews	USAF Mystic Star Channel
11121.0	USB		USAF MARS
11154.0	L/USB	Andrews	USAF Mystic Star Channel
11176.0	USB	Croughton	USAF GCCS
		Incirlik	USAF GCCS
		MAC Ops	MAC, Dhahran, Saudi Arabia
11191.0	USB		USN Tactical Channel
11204.0	USB	Architect	RAF Upavon
		Cyprus	RAF Akrotiri
11205.0	USB		USN Tactical Channel
11220.0	USB		Strategic Air Command
	L/USB	Andrews	USAF Mystic Star Channel
11222.0	USB	Stockholm	Stockholm Aero Radio
11226.0	USB		Strategic Air Command

11234.0	USB	Architect	RAF Upavon
		Cyprus	RAF Akrotiri
11243.0	USB		Strategic Air Command
			Royal Saudi Air Force
11271.0	USB	Lajes	USAF GCCS
11255.0	USB		USN HICOM Atlantic Ocean
11306.0	USB	Portishead	Portishead Aero Radio
11345.0	USB	Stockholm	Stockholm Aero Radio
11407.0	USB		USAF MARS
	L/USB	Andrews	USAF Mystic Star Channel
11466.0	L/USB	Andrews	USAF Mystic Star Channel
11484.0	L/USB	Andrews	USAF Mystic Star Channel
11488.0	L/USB	Andrews	USAF Mystic Star Channel
11494.0	USB		Strategic Air Command
11514.0	USB		Central Intelligence Agency
11538.0	USB		US Navy MARS
11605.0	AM		Kol Israel, Jerusalem
11965.0	AM		Voice of UAE, Abu Dhabi
12070.0	USB		Strategic Air Command
12085.0	AM		Radio Damascus
12133.0	USB	Portishead	Portishead Aero Radio
12168.0	USB	Portishead	Portishead Aero Radio
12215.0	USB		USN HICOM Indian Ocean
12230.0	USB		Central Intelligence Agency
12336.2	USB	YIR	Basrah Control, Iraq
12388.9	USB	JYO	Aqaba Radio, Jordan
12429.2	USB		International Distress Chan.
12658.2	CW	HZG	Damman Radio, Saudi Arabia
12660.0	CW	YIR	Basrah Control, Iraq
12709.0	CW	A9M	Bahrain Radio, Bahrain
12725.0	CW	JYO	Aqaba Radio, Jordan
12747.0	USB	MIW2	MOSSAD
12950.0	USB	SYN2	MOSSAD
	USB	VLB2	MOSSAD
13100.8	USB	Portishead	Portishead Maritime Radio
13107.0	USB	YIR	Basrah Control, Iraq
13116.3	USB	Portishead	Portishead Maritime Radio
13159.7	USB	JYO	Aqaba Radio, Jordan
13172.1	USB	Portishead	Portishead Maritime Radio
13196.9	USB	Portishead	Portishead Maritime Radio
13205.0	L/USB	Andrews	USAF Mystic Star Channel
13211.0	USB		Strategic Air Command
13214.0	USB	Croughton	USAF GCCS
		Lajes	USAF GCCS

		Lajes	USAF GCCS
13215.0	USB	Lajes	USAF GCCS
13217.0	L/USB	Andrews	USAF Mystic Star Channel
13241.0	USB		Strategic Air Command
13244.0	USB	Lajes	USAF GCCS
13247.0	USB		Strategic Air Command
13288.0	USB		Gulf ATC
13312.0	USB		Gulf ATC
13324.0	USB	Berne	Berne Aero Radio
13333.0	USB	Speedbird	British Airways Company Chan.
13339.0	USB	Jeddah	Saudia Company Channel
	USB	Falcon	Gulf Air Company Channel
13342.0	USB	Jeddah	Saudia Company Channel
	USB	Stockholm	Stockholm Aero Radio
13415.0	USB		Central Intelligence Agency
13576.0	USB	Stockholm	Stockholm Aero Radio
13605.0	AM		Voice of UAE, Abu Dhabi
13610.0	AM		Radio Kuwait
13620.0	AM		Radio Kuwait
13643.0	USB		US Navy MARS
13655.0	AM		Radio Jordan
13660.0	AM		Radio Baghdad
13714.0	USB		US Army MARS
13741.0	USB		US Army MARS
13823.0	L/USB	Andrews	USAF Mystic Star Channel
13827.5	USB		US Navy MARS
13865.0	USB	Portishead	Portishead Aero Radio
13878.0	L/USB	Andrews	USAF Mystic Star Channel
13921.0	USB	CIO2	MOSSAD
13927.0	USB		USAF MARS
13942.0	USB	Stockholm	Stockholm Aero Radio
13974.0	USB		US Navy MARS
13977.0	USB		USAF MARS
13992.0	USB		US Army MARS
13993.0	USB		USAF/US Army MARS
14355.0	USB		Central Intelligence Agency
14380.0	L/USB		PLO
14385.0	USB		US Navy MARS
14388.0	L/USB		PLO
14390.0	L/USB		PLO
14390.5	USB		USAF MARS
14392.0	USB		USAF MARS
14393.0	L/USB		PLO
14395.0	L/USB		PLO

14399.0	L/USB		PLO
14402.0	USB		US Army MARS
14404.0	USB		US Army MARS
14405.0	USB		USAF MARS
14408.0	USB		US Army/Air Force MARS
14441.5	USB		USN MARS Calling Channel
14448.0	USB		Central Intelligence Agency
14463.5	USB		US Navy MARS
14470.0	USB		US Navy MARS
14477.0	USB		US Navy MARS
14483.5	USB		US Navy MARS
14485.0	USB		US Army MARS
14487.0	USB		US Army MARS
14510.0	USB		US Army MARS
14511.0	USB		US Army MARS
14512.0	USB		US Army MARS
14579.0	USB		US Army MARS
14580.0	USB		US Army MARS
14606.0	USB		USAF MARS
14607.5	USB		USAF MARS
14633.5	USB		US Army MARS
14666.0	USB		US Army MARS
14703.0	USB		Central Intelligence Agency
14750.0	USB	VLB2	MOSSAD
14752.0	USB		Central Intelligence Agency
14811.0	USB		Central Intelligence Agency
14818.5	USB		US Navy MARS
14829.0	L/USB	Andrews	USAF Mystic Star Channel
14840.0	USB		US Navy MARS
14890.0	USB	Portishead	Portishead Aero Radio
14935.5	USB		US Navy MARS
15015.0	USB	Incirlik	USAF GCCS
	L/USB	Andrews	USAF Mystic Star Channel
15021.0	USB	Stockholm	Stockholm Aero Radio
15036.0	USB	Croughton	USAF GCCS
		Incirlik	USAF GCCS
15044.0	USB		Strategic Air Command
15046.0	USB	Berne	Berne Aero Radio
15095.0	AM		Radio Damascus
15320.0	AM		UAE Radio, Dubai
15435.0	AM		UAE Radio, Dubai
15495.0	AM		Radio Kuwait
15505.0	AM		Radio Kuwait
15640.0	AM		Kol Israel, Jerusalem

15680.0	L/USB	Andrews	USAF Mystic Star Channel
15685.0	USB	Stockholm	Stockholm Aero Radio
15835.0	USB	Berne	Berne Aero Radio
15938.0	USB		Central Intelligence Agency
16300.0	USB		US Navy MARS
16370.0	USB	Portishead	Portishead Aero Radio
16377.0	USB		Central Intelligence Agency
16387.5	USB		US Navy MARS
16522.0	USB		International Distress Chan.
16537.5	USB	YIR	Basrah Control, Iraq
16860.8	CW	HZG	Damman Radio, Saudi Arabia
16880.0	CW	YIR	Basrah Control, Iraq
16906.0	CW	YIR	Basrah Control, Iraq
17169.0	CW	A9M	Bahrain Radio, Bahrain
17170.0	USB	MIW2	MOSSAD
17175.2	CW	A9M	Bahrain Radio, Bahrain
17263.9	USB	Portishead	Portishead Maritime Radio
17276.3	USB	Portishead	Portishead Maritime Radio
17285.6	USB	Portishead	Portishead Maritime Radio
17310.4	USB	YIR	Basrah Control, Iraq
17344.5	USB	Portishead	Portishead Maritime Radio
17405.0	USB	Portishead	Portishead Aero Radio
17487.0	USB		USAF MARS
17710.0	AM		Radio Damascus
17865.0	AM		UAE Radio, Dubai
17916.0	USB	Stockholm	Stockholm Aero Radio
17922.0	USB	Speedbird	British Airways Company Chan.
17925.0	USB	Jeddah	Saudia Company Channel
17931.0	USB	Falcon	Gulf Air Company Channel
17966.0	USB	CIO2	MOSSAD
17972.0	USB		Strategic Air Command
	L/USB	Andrews	USAF Mystic Star Channel
17975.0	USB		Strategic Air Command
17982.0	USB	Stockholm	Stockholm Aero Radio
18018.0	USB	Architect	RAF Upavon
		Cyprus	RAF Akrotiri
18023.0	USB	Berne	Berne Aero Radio
18175.0	L/USB	Andrews	USAF Mystic Star Channel
18178.0	USB	VLB2	MOSSAD
18210.0	USB	Portishead	Portishead Aero Radio
18320.0	L/USB	Andrews	USAF Mystic Star Channel
18480.0	USB		Central Intelligence Agency
19510.0	USB	Portishead	Portishead Aero Radio
20065.0	USB	Portishead	Portishead Aero Radio

20115.0	USB		Central Intelligence Agency
20167.0	USB		Strategic Air Command
20188.5	USB		USAF MARS
20524.0	USB		Central Intelligence Agency
20678.5	USB		US Navy MARS
20680.0	USB		US Navy MARS
20737.0	USB		USAF MARS
20740.0	USB	MIW2	MOSSAD
20770.0	USB	Stockholm	Stockholm Aero Radio
20811.0	USB		US Army MARS
20870.0	USB	Berne	Berne Aero Radio
20936.0	USB		US Navy MARS
20937.5	USB		US Navy MARS
20940.0	USB		US Army MARS
20988.5	USB		US Navy MARS
20991.0	USB		US Army MARS
20992.5	USB		USAF MARS
20993.0	USB		US Army MARS
20994.0	USB		USAF/US Army MARS
21410.0	USB		Kuwaiti Air Force
21415.0	USB		Kuwaiti Air Force
21605.0	AM		UAE Radio, Dubai
21765.0	USB	Portishead	Portishead Aero Radio
21943.0	USB	Falcon	Gulf Air Company Channel
21946.0	USB	Speedbird	British Airways Company Chan.
21988.0	USB	Berne	Berne Aero Radio
21994.0	USB	Jeddah	Saudia Company Channel
21997.0	USB	Stockholm	Stockholm Aero Radio
22037.2	USB	YIR	Basrah Control, Iraq
22322.0	CW	A9M	Bahrain Radio, Bahrain
22338.0	CW	YIR	Basrah Control, Iraq
22338.2	CW	HZG	Damman Radio, Saudi Arabia
22633.2	USB	YIR	Basrah Control, Iraq
23035.0	L/USB	Andrews	USAF Mystic Star Channel
23195.0	USB	CIO2	MOSSAD
23210.0	USB	Stockholm	Stockholm Aero Radio
23220.0	USB	Architect	RAF Upavon
23227.0	USB	Incirlik	USAF GCCS
23285.0	USB	Berne	Berne Aero Radio
23287.0	USB		USN HICOM Atlantic Ocean
23307.0	USB	Jeddah	Saudia Company Channel
23315.0	USB		USN HICOM Indian Ocean
23385.0	L/USB	Andrews	USAF Mystic Star Channel
25500.0	USB	Berne	Berne Aero Radio

READING LIST

If you wish to find out more about monitoring the Iraq/Kuwait conflict and obtain more frequencies and information, the following reading list is recommended.

Maritime Radio Handbook by M. Schaay lists the world's coastal stations and there is a section covering the Middle East. A must for monitoring shipping. Price £8.95.

Aeronautical Radio Communications by R. E. Evans shows how to monitor both military and civilian aircraft and takes you step by step through this fascinating subject. Also included is background information, transmission modes, propagation and a large frequency list which has been monitored by the author. Price £18.95.

Aeronautical Radio Handbook by M. Schaay lists hundreds of aeronautical frequencies worldwide used by the military, civilian and meteorological stations. Price £8.75.

U.S. Military Radio Communications by M. Schaay will be useful for eavesdropping on all U.S. military HF communications, including Army, Navy and Air Force. Hundreds of frequencies are listed, but only Part 1 is recommended because it covers the Gulf. Price £12.50.

Passport to World Band Radio 1991 covers the broadcasting stations throughout the world. So if you want to hear the news or comments from the Gulf states this will be an ideal reference, and is set out in a graphic format for quick reference. It also lists English language broadcasts by country. Price £13.50.

Monitoring Times is a monthly US radio magazines which has many articles with frequencies on military, aviation and utility stations not found in any British magazine. Price £2.85 per single copy or £27 for annual subscription.

All these books are obtainable from **Interbooks, 8 Abbot Street, Perth, PH2 0EB, Scotland. Tel: 0738-30707** In the UK their postal charges are £1.25 for 1 book, £1.75 for 2, £2.25 for 3, £2.75 for 4, and £3 for 5 or more. It is best to enquire the overseas postage rates.